AF342306

# THE ANARCHIST HANDBOOK 3

## ROBERT WELLS

J. FLORES PUBLICATIONS
P.O. BOX 830131
MIAMI, FL 33283-0131

# WARNING NOTICE!

Neither the author nor the publisher assumes any responsibility for the use or misuse of the information contained in this book. The author and publisher specifically disclaim any personal liability, loss, or risk incurred as a consequence of the use and application, either directly or indirectly, of any advice or information presented herein.

Be advised that there may be certain items represented in this book as to which the sale, possession, construction or interstate transportation thereof may be restricted, prohibited or subject to special licensing requirements. Consult with your local ATF and law enforcement authorities in your area before obtaining or constructing such items.

Technical data presented herein, particularly technical data on dealing with dangerous chemicals, poisons, explosives, drugs, and safety procedures inevitably reflects the author's individual beliefs and experience with particular equipment and environments under specific circumstances which the reader cannot duplicate or experience exactly. The information in this book should therefore be used for guidance only and should be approached with great caution.

Since neither the author nor the publisher have control over the materials used, use of techniques or equipment, or the abilities of the reader, no responsibility, either implied or expressed, is assumed for the use or misuse of the data or procedures given in this book.

# TABLE OF CONTENTS

# INTRODUCTION

For quite a few years now, considerable amounts of time, effort, and money have been expended on improving older sabotage devices and accessory gear and in developing new and better items. As a result, a wide variety of manufactured explosive and incendiary items are available for use.

At the same time a somewhat smaller effort has been devoted to improvising and testing homemade or field expedient devices and techniques for accomplishing similar results.

Since the manufactured, precision devices almost always will be more effective, more reliable, and easier to use, why spend time on improving field expedients to do the job?

For one thing, shelf items will just not be available for certain operations for security or logistical reasons. In these cases the operator will have to rely on materials he can buy in a drug or paint store, find in a junk pile, or scrounge from military stocks.

Secondly, many of the ingredients and

materials used in fabricating homemade items are so commonplace or innocuous they can be carried without arousing suspicion. The completed item itself often is more easily concealed or camouflaged.

In addition, the field expedient item can be tailored for the intended target, thereby providing an advantage over the standard item in flexibility and versatility.

While most of the pertinent information on sabotage shelf items is available in catalogues and other publications, much of the improvisation know-how has remained in the minds or files of a few individuals.

It is the intent of this manual to consolidate and bring up to date a selected body of this information and make it available for wider use.

The devices and techniques included have been selected because they have been well proved out and because they are practical, versatile, and not overly difficult to construct. Other techniques have been omitted because they are unreliable or are just too hazardous to attempt except under the most carefully controlled conditions.

The techniques described are not so complex as to require a chemical laboratory or machine shop; however, many of them do assume access to a few household tools.

# CHAPTER 1
# EXPLOSIVES AND INCENDIARIES

**HOW TO MAKE SAFETY FUSE**

This item consists of string, twine, or shoelaces that have been treated with either a mixture of potassium nitrate and granulated sugar or potassium chlorate and granulated sugar.

Depending upon the length of the fuse, the user can be away from the immediate scene when an incendiary system is initiated.

**Material And Equipment Required:**

String, twine or shoelaces made of cotton or linen.

Potassium nitrate or potassium chlorate.

Granulated sugar.

Small cooking pot.

Spoon.

Heat source such as stove or hot plate.

Soap.

## Preparation:

1. Wash string or shoelaces in hot soapy water; rinse in fresh water.

2. Dissolve one part potassium nitrate or potassium chlorate and one part granulated sugar in two parts hot water.

3. Soak string or shoelaces in the hot solution for at least five minutes.

4. Remove the string from hot solution and twist or braid three strands or string together.

5. Hang the fuse up to dry.

6. Check actual burning rate of the fuse by measuring the time it takes for a known length to burn.

## How To Use:

1. This fuse does not have a waterproof coating and it must be tested by burning a measured length before actual use.

2. Cut the fuse long enough to allow a reasonable time delay in initiation of the incendiary system.

3. Insert one end of the fuse in a quantity of an igniter mixture so that the fuse end terminates near the center of the mixture. Be sure the fuse cord is anchored in the igniter mixture and

cannot pull away. In the case of a solid igniter material, the improvised string fuse is securely wrapped around a piece of solid igniter material.

4. The fuse is initiated by lighting the free end of the fuse with a match.

5. This fuse does not burn when it is wet. Its use is not recommended where there is the possibility of the fuse getting wet.

## MATCH HEAD IGNITER

A good ignition material for incendiaries can be obtained from the heads of safety matches, which are available almost any place. The composition must be removed from the heads of many of them to get a sufficient quantity of igniter material. It will ignite napalm, wax and sawdust, paper, and other flammables.

### Materials:
Safety matches
Knife or pliers
Moisture-tight container

### Preparation:
1. Remove the match head composition by scraping with a knife or crushing with pliers. Collect several spoonfuls of it and store in a moisture-tight container.

2. Put at least 2 spoonfuls on the material to be ignited. To ignite liquids, such as solvents or napalm, wrap several spoonfuls in a piece of paper and hang this just over the fluid, or place nearby. If fluids dampen the mixture it may not ignite.

Ignition can be by time fuse, firecracker fuse, a spark, or concentrated sulfuric acid.

## PARAFFIN-SAWDUST INCENDIARY

Paraffin-sawdust is almost as effective as napalm against combustible targets, but it is slower in starting. It is solid when cool and thus is more easily carried and used than liquid napalm. In addition, it can be stored indefinitely without special care.

### Materials:
Dry sawdust
Paraffin, beeswax, or candle wax
Heat source
Container

### Preparation:
1. Melt the wax, remove the container from the fire and stir in a roughly equal amount of sawdust.

2. Continue to stir the cooling mixture until it becomes almost solid, then remove from the container and let it cool and solidify further.

Lumps of the mixture the size of a fist are easiest to manage. The chunks of incendiary may be carried to the target in a paper bag or other wrapper. Any igniter that will set fire to the paper wrapper will ignite the wax and sawdust.

A similar incendiary can be made by dipping sheets of newspaper into melted wax and allowing them to cool. These papers may then be crumpled up and used in the same manner as the paraffin-sawdust, although they will not burn as hot and persistently.

## MATCH HEAD PIPE BOMB

Improvised hand grenades can be made from a piece of iron pipe and matchheads. A Similar non-metalic bomb can be constructed be subsituding the iron pipe and pipe caps with PVC tubing and end caps. Hot glue the end caps to the tube using a hot melt glue gun.

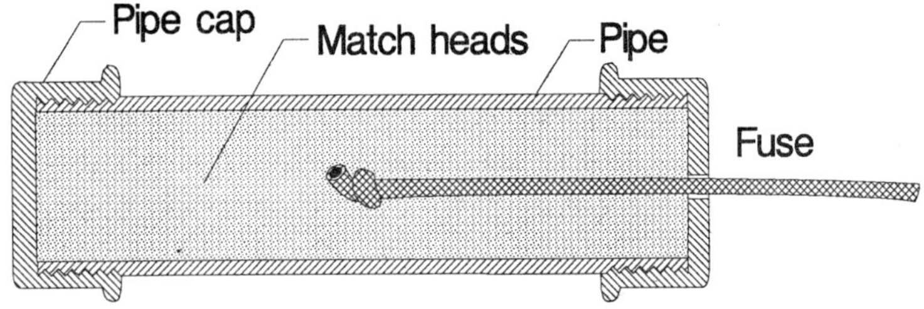


### Material Required:
Iron pipe, threaded ends, 1½ inch to 3 inch diameter, 3 inch to 8 inches long.
Two (2) iron pipe caps.
Matchs
Fuse Cord.
Hand Drill.

### Preparation:
1. Make a knot at one end of a piece of fuse.

Knot

*NOTE: To find out how long the fuse cord should be, check the time it takes a known length to burn. If 12 inches burns in 30 seconds, a 6-inch cord will ignite the grenade in 15 seconds.*

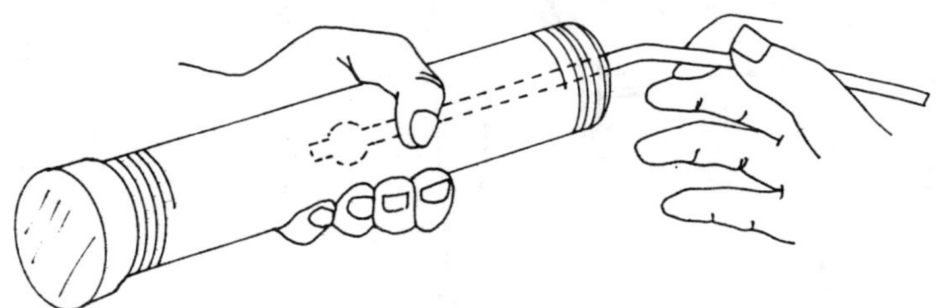

2. Screw pipe cap to one end of pipe. Place fuse cord into the opposite end so that the knot is near the center of the pipe.

3. Pour the matchheads into pipe a little bit at a time. Tap the base of the pipe frequently to settle filler.

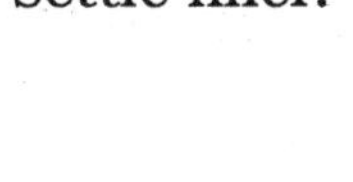

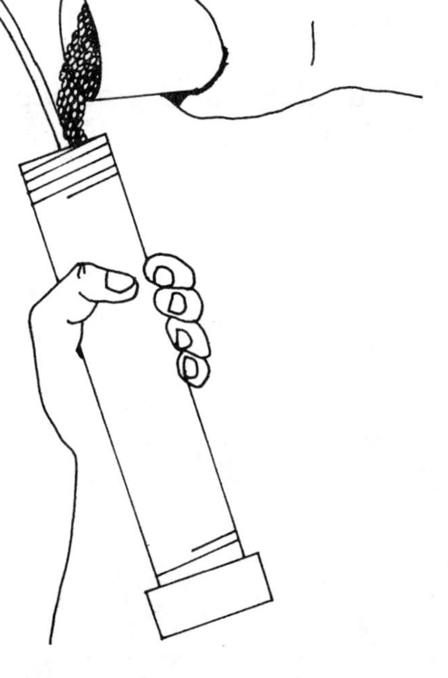

4. Drill a hole in the center of the unassembled pipe cap large enough for the fuse cord to pass through.

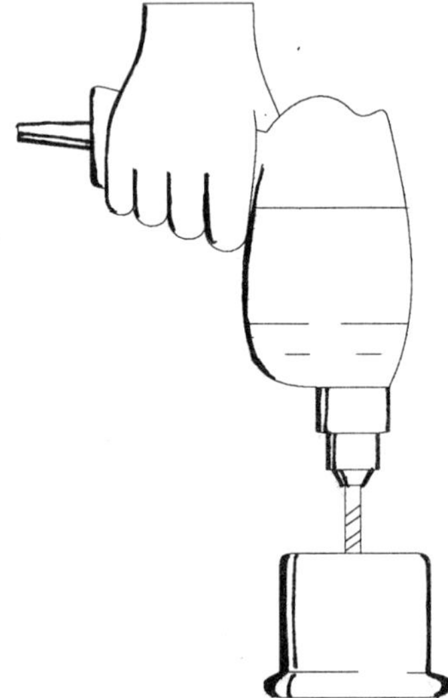

5. Wipe pipe threads to remove any filler material.

6. Slide the drilled pipe cap over the fuse and screw hand tight onto the pipe.

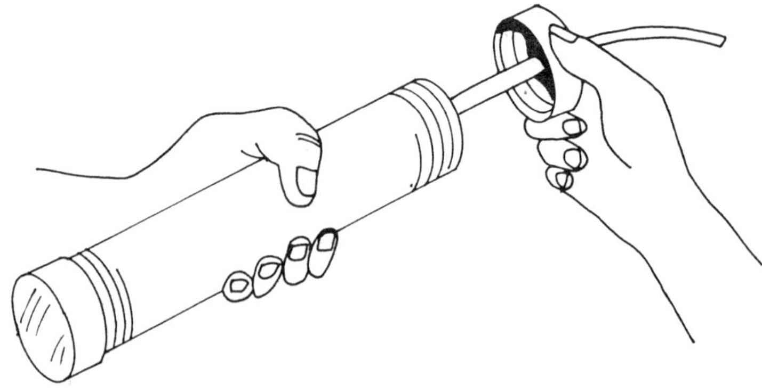

# MECHANICALLY INITIATED FIRE BOTTLE

The mechanically initiated Fire Bottle is an incendiary device which ignites when thrown against a hard surface.

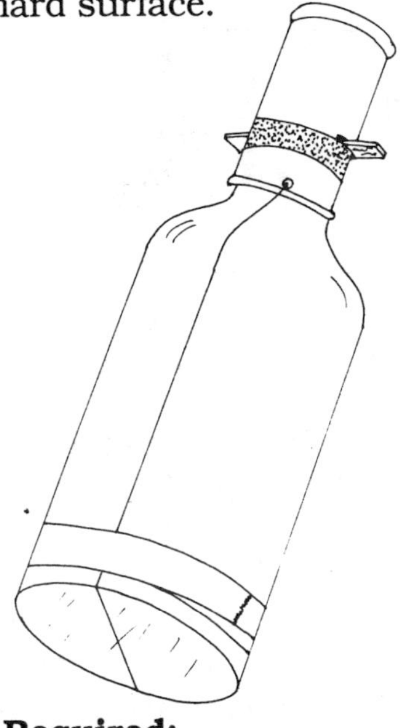

## Materials Required:

Glass jar or short neck bottle with a leakproof lid or stopper.

"Tin" can or similar container just large enough to fit over the lid of the jar.

Coil spring (compression) approximately ½ the diameter of the can and 1 ½ times as long.

Gasoline.

Four (4) "blue tip" matches.

Flat stick or piece of metal (roughly ½ inch x ¹⁄16 inch x 4 inches)

Wire or heavy twine.

Adhesive tape.

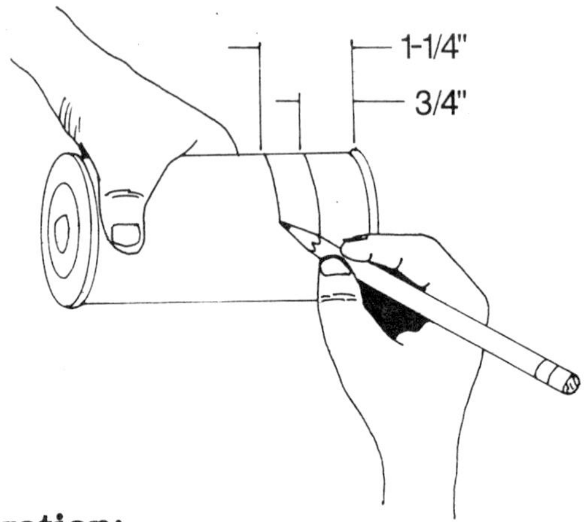



## Preparation:

1. Draw or scratch two lines around the can—one ¾ inch and the other 1¼ inch from the open end.

2. Cut 2 slots on opposite sides of the can at the line farthest from the open end. Make slots large enough for the flat stick or piece of metal to pass through.

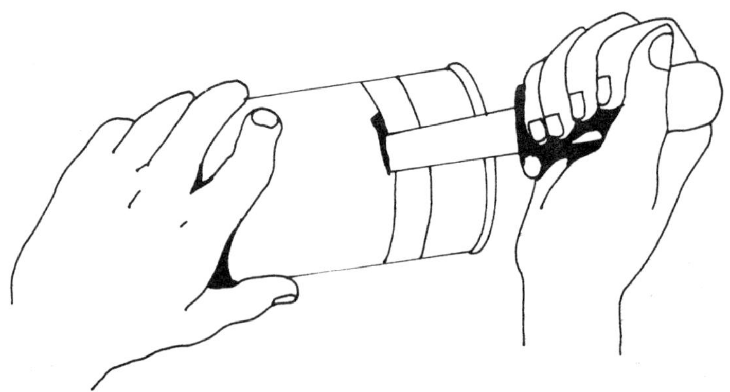

3. Punch 2 small holes just below the rim of the open end of the can.

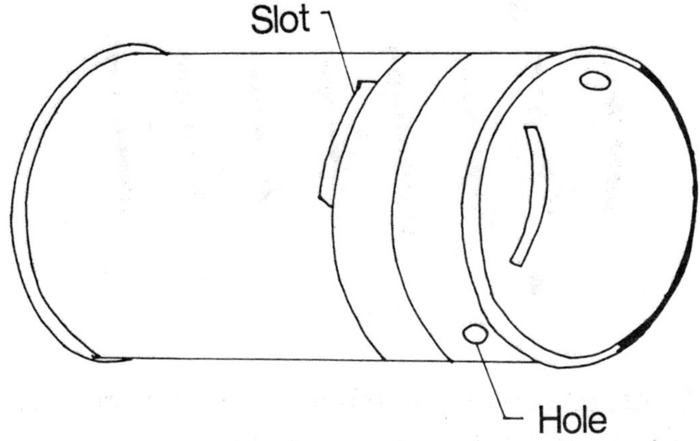


4. Tape blue tip matches together in pairs. The distance between the match heads should equal the inside diameter of the can. Two pairs are sufficient.

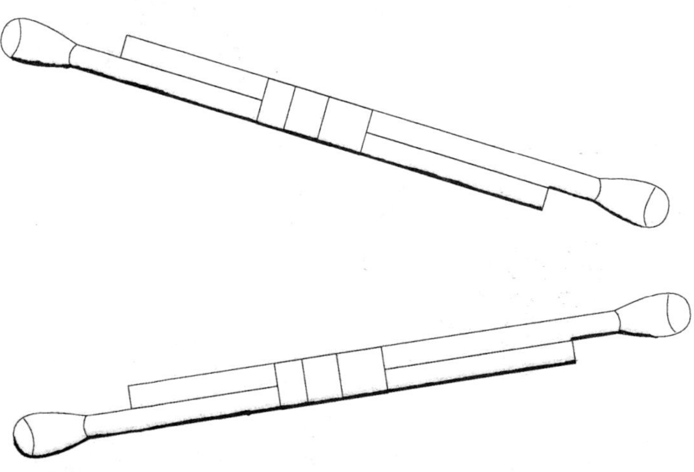

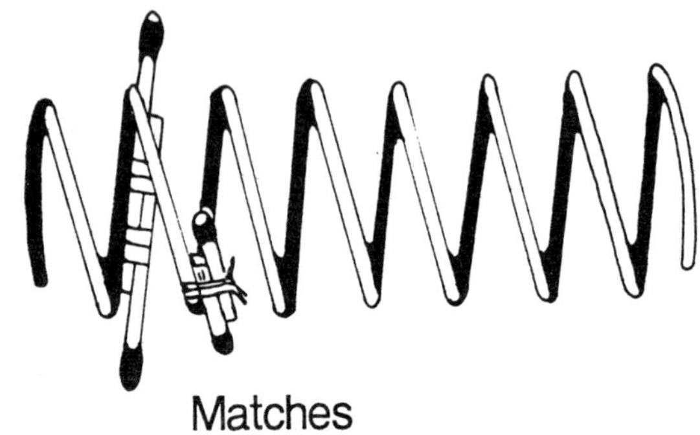

Matches

5. Attach paired matches to second and third coils of the spring, using thin wire.

6. Insert the end of the spring opposite the matches into the tin can.

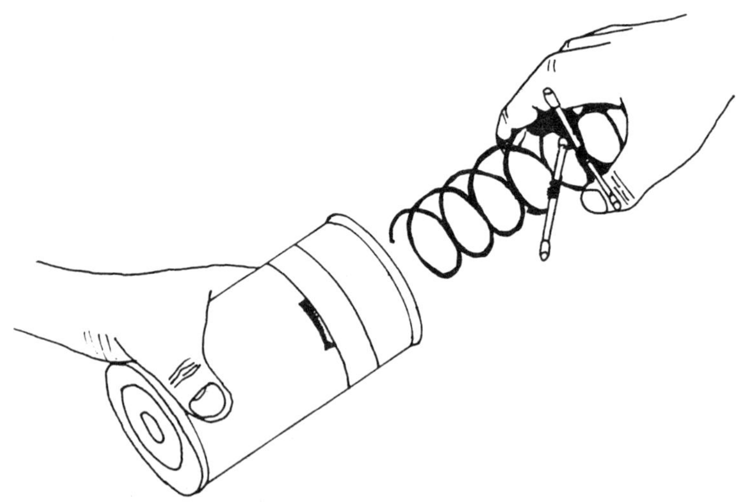

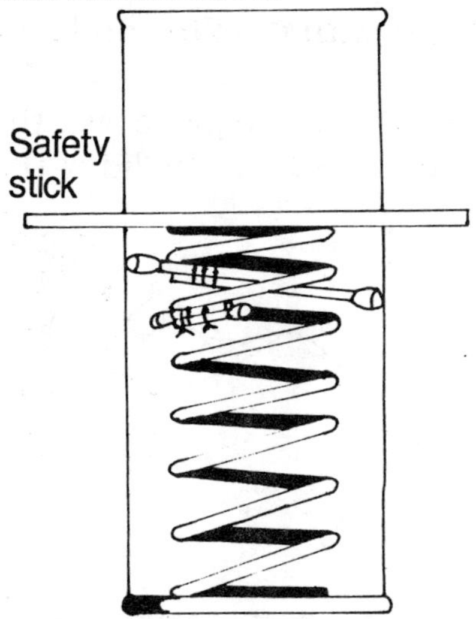



7. Compress the spring until the end with the matches passes the slot in the can. Pass the flat stick or piece of metal through slots in can to hold spring in place. This acts as a safety device.

8. Punch many closely-spaced small holes between the lines marked on the can to form a striking surface for the matches. Be careful not to seriously deform can.

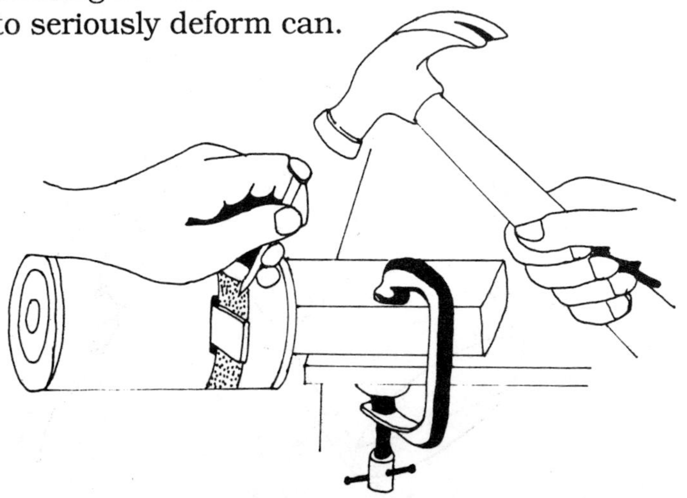

9. Fill the jar with gasoline and cap tightly.

10. Turn can over and place over the jar so that the safety stick rests on the lid of the jar.

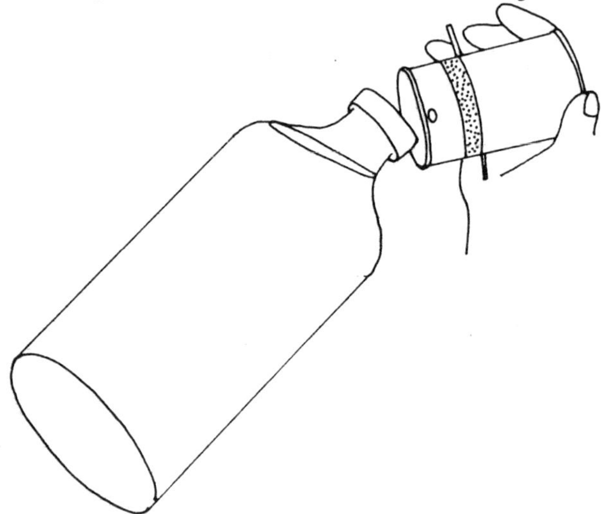

11. Pass wire or twine around the bottom of the jar. Thread ends through holes in can and bind tightly to jar.

12. Tape wire or cord to jar near the bottom.

## How To Use:

1. Carefully withdraw flat safety stick.

2. Throw jar at hard surface.

## CAUTION: DO NOT REMOVE SAFETY STICK UNTIL READY TO THROW FIRE BOTTLE.

The safety stick, when in place, prevents ignition of the fire bottle if it should accidentally be broken.

# CHAPTER 2
# IGNITER AND DELAY MECHANISMS

**CANDLE**

This delay ignites flammable fuels of low volatility such as fuel oil and kerosene. A lighted candle properly inserted in a small container of flammable liquid of low volatility causes ignition of the flammable liquid when the flame burns down to the liquid level. The flame from the

burning liquid is used to ignite incendiary material such as paper, straw, rags, and wooden structures. The delay time is reasonably accurate, and may be easily calibrated by determining the burning rate of the candle. No special skills are required to use this delay. Shielding is required for the candle when used in an area of strong winds or drafts. This delay is not recommended for use with highly volatile liquids because premature ignition may take place. This device is useful where a delay of one hour or longer is desired. The candle delay works well in cold or hot weather, and has the advantage of being consumed in the resulting fire, thus reducing evidence of arson.

**Material And Equipment Required:**
Candle.
Bowl.
Perforated can or carton.
Fuel oil or kerosene.
Matches.
Small piece of cloth.

**Preparation:**
1. Make two marks on the side of the candle, 1 ½ inches and 2 inches from the top. Light the candle and record the times at which the wax melts at the marks on the side.

2. The distance burned by the candle divided by the elapsed time determines the burning rate of the candle.

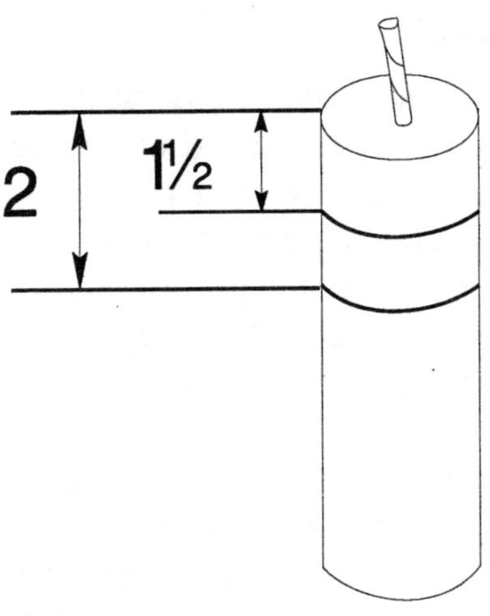


## APPLICATION

1. Using a lighted candle of desired length, drip hot wax in the center of the bowl. Melt the base of the candle with a lighted match. Firmly press the softened base of the candle into the hot wax in the center of the bowl. Be sure the candle will stand up securely without toppling over. Extinguish the candle. Wrap a small piece of cloth around the candle and slide it down to the bottom of the bowl. Place a quantity of fuel oil or kerosene in the bowl. Be sure that the level of the fluid reaches the cloth, so it will act as a wick. Pile the incendiary material around the bowl where it can catch fire after the fuel oil or kerosene ignites.

2. If this delay must be set in a windy or drafty location, place a shield over it. Notch or punch holes in a metal can or cardboard carton at the bottom and sides for ventilation, and place this cover over the delay.

## CIGARETTE

This item consists of a bundle of matches wrapped around a lighted cigarette. It is placed directly on easily ignited material. Ignition occurs when the lighted portion of the burning cigarette reaches the match heads. A cigarette delay directly ignites the following incendiaries: Napalm, and Gelled Gasoline (improvised thickeners).

The following dry tinder type materials may also be directly ignited by the cigarette delay mechanism: Straw, paper, hay, wood shavings and rags.

Usually this delay will ignite in fifteen (15) to twenty (20) minutes, depending on length of cigarette, make of cigarette, and force of air currents. A duplicate delay mechanism should be tested to determine delay time for various ambient conditions.

The cigarette must be placed so that the flame will travel horizontally or upward. A burning cigarette that is clamped or held will not burn past the point of confinement. Therefore, the cigarette should not contact any object other than the matches.

## Materials And Equipment Required:

Cigarette.
Matches (wooden).
Match box.
String or tape.

## Preparation:

For Picket-fence delay

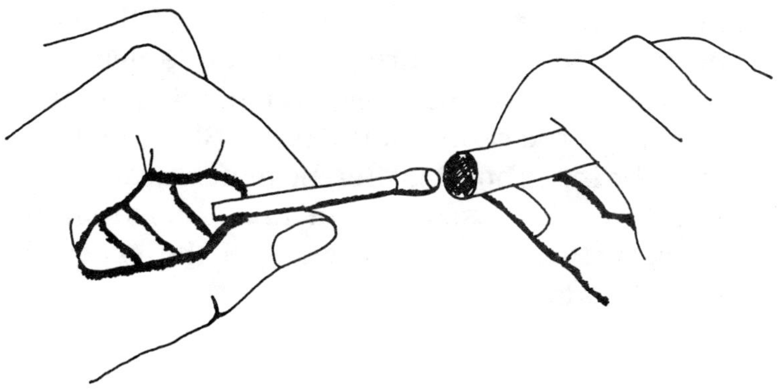

1. Push one wooden match head into a cigarette a predetermined distance to obtain the approximate delay time.

2. Tie or tape matches around the cigarette with the match heads at the same location as the first match in the cigarette.

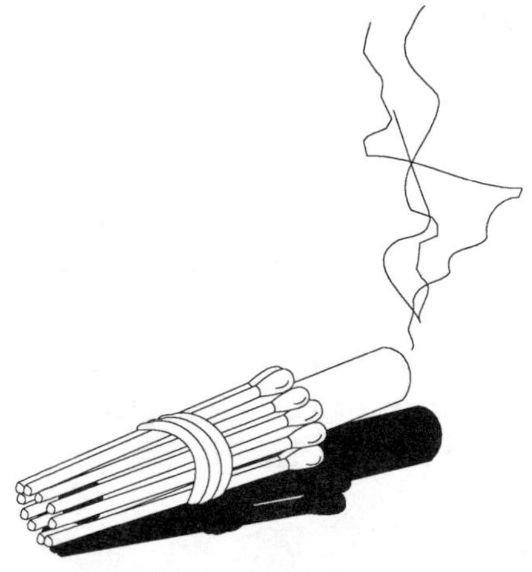

For Match box delay

Tear out one end of the inner tray of a box of matches (the end next to the match holds). Push one match into the cigarette, insert this cigarette into the book of matches and parallel to the matches at the center of the pack. Slide the tray out of the inner box, leaving the match heads and the cigarette exposed. The head of the match in the cigarette should be even with the exposed match heads.

## How To Use:
Picket-fence delay.

1. Light the cigarette and place the delay mechanism on a pile of igniter mixture, paper, straw, or other dry tinder type material. Be sure that the portion of the cigarette between the lid end and the match heads is not touching any-thing.

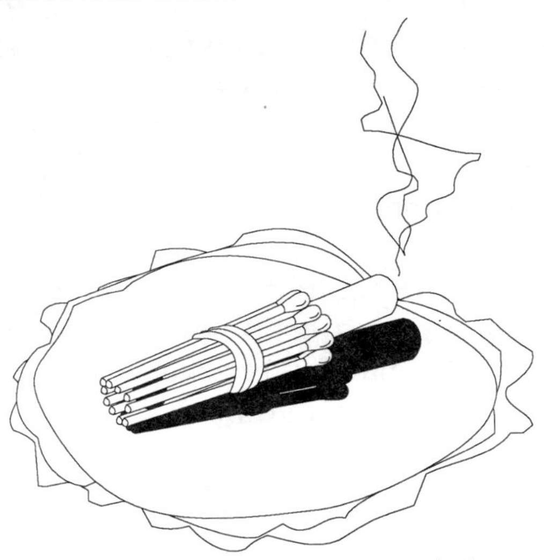

2. Pile tinder material all around the cigarette to enhance ignition when the match heads ignite.

Match box delay

1. Place the delay so that the cigarette is horizontal and on top of the material to be ignited. Light the cigarette.

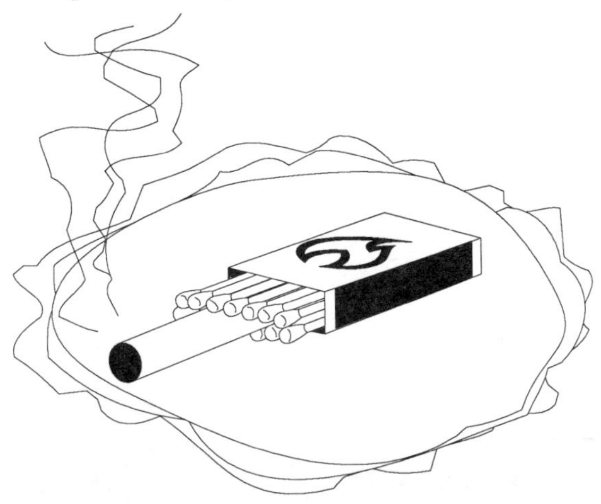

2. Be sure ignitable material such as paper, straw, flammable solvents, or napalm is placed close to the match heads. When using flammable solvents, light the cigarette away from the area of solvent fumes.

3. To assure ignition of the target, sprinkle some igniter material on the combustible material. The match box delay is then placed on top of the igniter material.

# DELAY IGNITER FROM CIGARETTE

A simple and economical time delay can be made with a common cigarette.

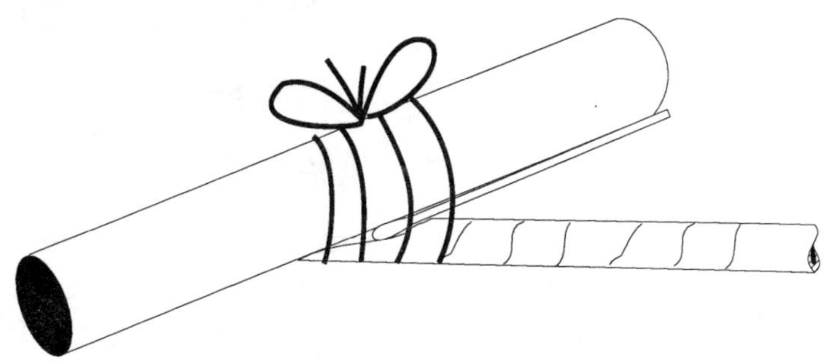

## Materials Required

Cigarette
Paper match
String (shoelace or similar cord)
Fuse (improvised or commercial)

## Preparation

1. Cut end of fuse to expose inner core.

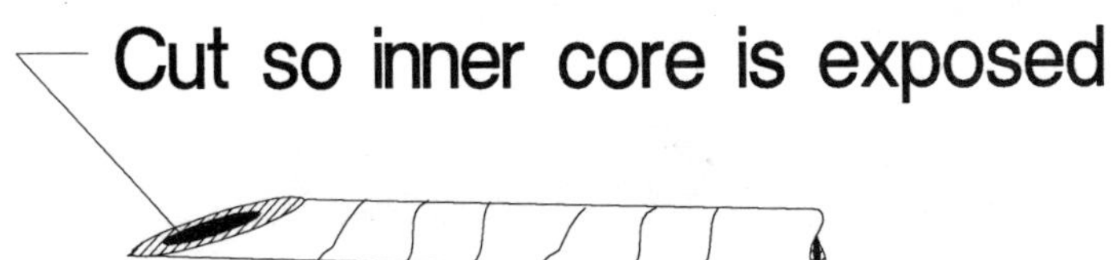

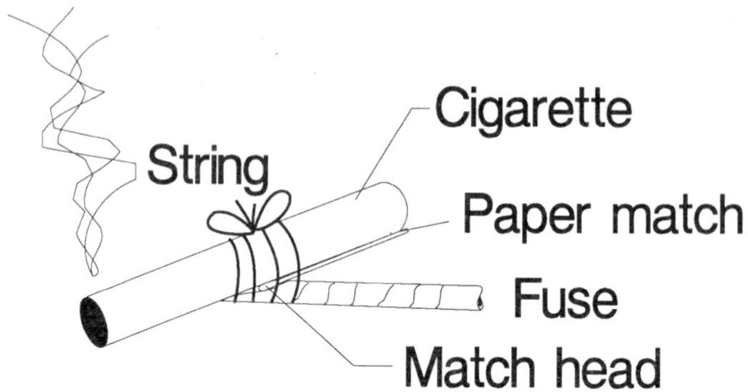


2. Light cigarette in normal fashion. Place a paper match so that the head is over exposed end of fuse cord and tie both to the side of the burning cigarette with string.

3. Position the burning cigarette with fuse so that it burns freely. A suggested method is to hang the delay on a twig.

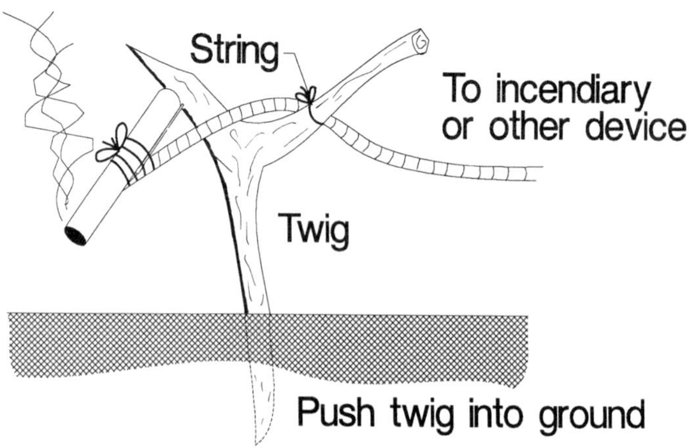

*NOTE: Common dry cigarettes burn about one inch every seven or eight minutes in still air. If the fuse cord is placed one inch from the burning end of a cigarette, a time delay of seven or eight minutes will result.*

Delay time will vary depending upon type of cigarette, wind, moisture, and other atmospheric conditions.

To obtain accurate delay time, a test run should be made under "use" conditions.

## FUSE IGNITER FROM BOOK MATCHES

A simple, reliable fuse igniter can be made from paper book matches.

### Material Required:

Paper book matches.
Adhesive or friction tape.
Fuse cord (improvised or commercial).
Pin or small nail.

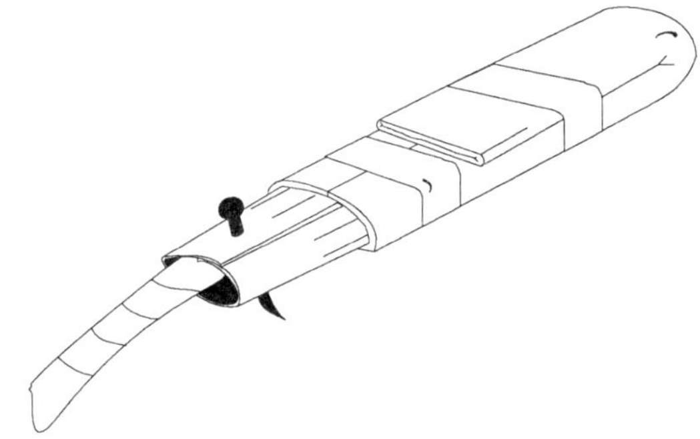

### Preparation:

1. Remove the staple(s) from match book and separate matches from cover.

2. Cut fuse cord so that inner core is exposed. (If using commercial fuse cord.)

3. Tape exposed end of fuse cord in center of one row of matches.

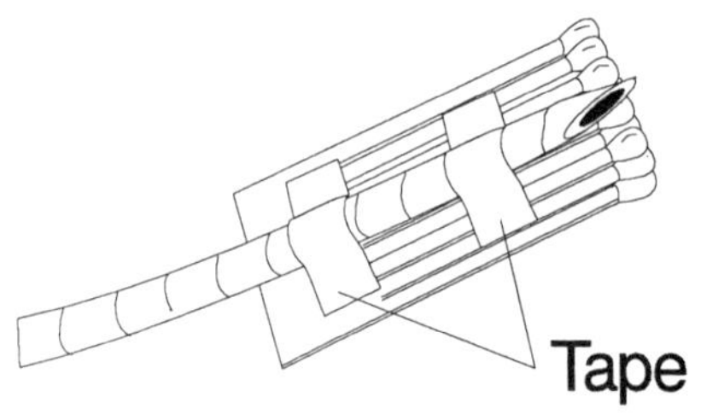

4. Fold matches over fuse and tape.

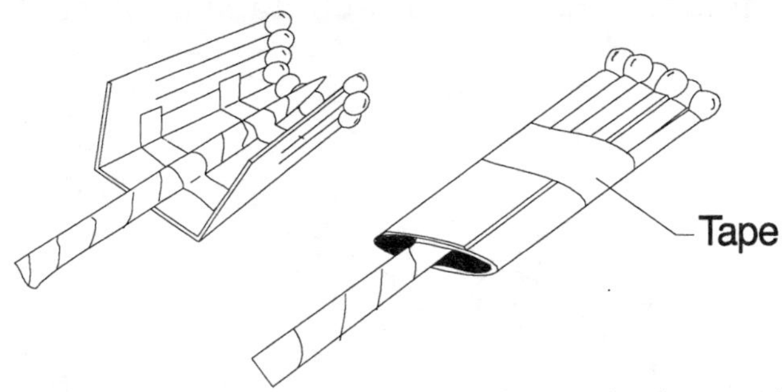



5. Shape the cover into a tube with the striking surface on the inside and tape. Make sure the edges of the cover at the striking end are butted. Leave cover open at opposite end for insertion of the matches.

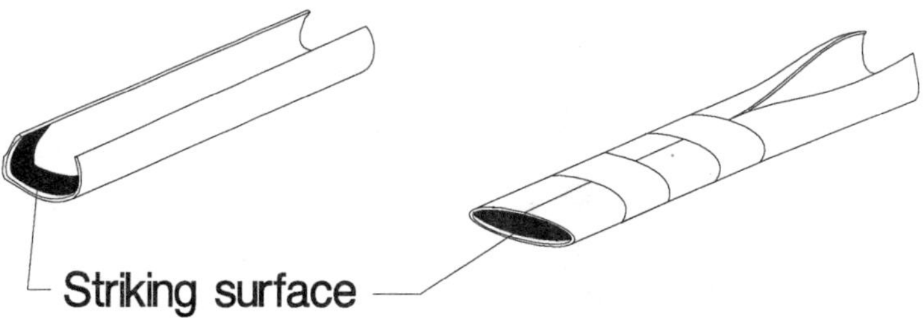



6. Push the taped matches with fuse cord into the tube until the bottom ends of the matches are exposed about ¾ inch.

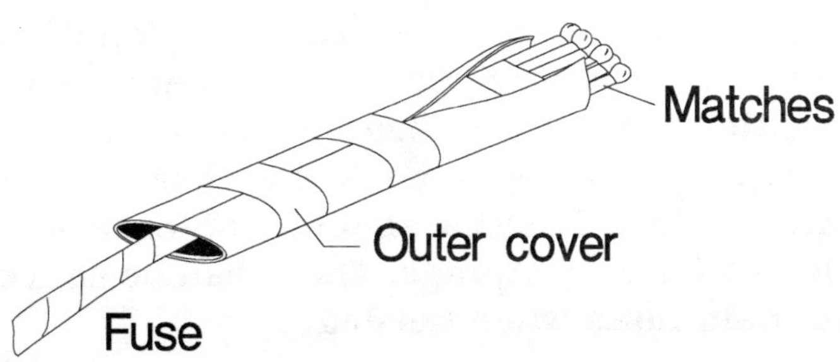

7. Flatten and fold the open end of the tube so that it laps over about 1 inch; tape in place.

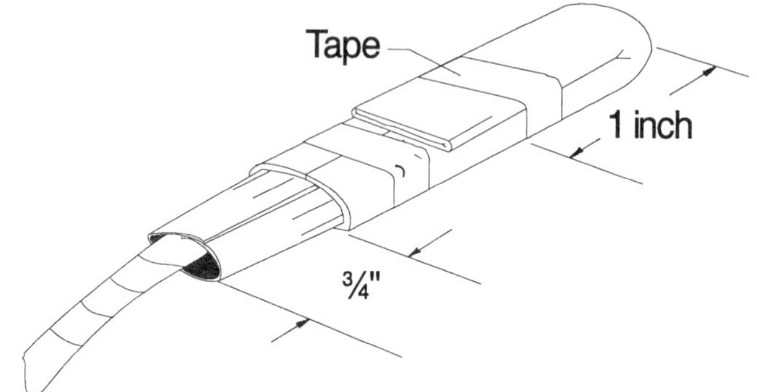


8. Push pin or small nail through matches and fuse cord. Bend end of pin or nail.

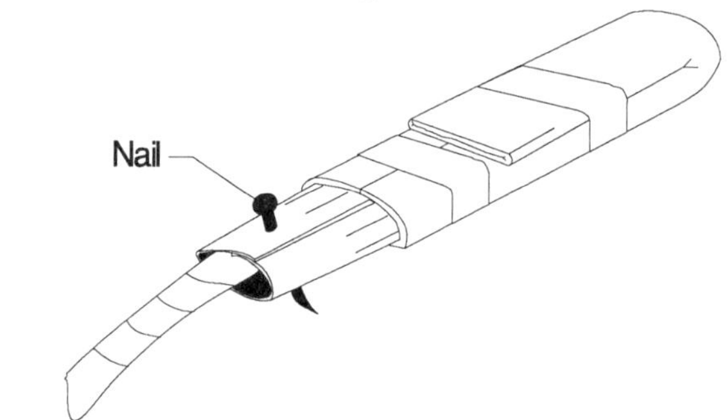


## How To Use:

To light the fuse cord, the igniter is held by both hands and pulled sharply or quickly.

**CAUTION: Store matches and completed fuse igniters in moisture-proof containers such as plastic or rubber type bags until ready for use. Damp or wet paper book matches will not ignite. Fuse lengths should not exceed 12 inches for easy storage. These can be spliced to main fuses when needed.**

# CHAPTER 3
# BOOBYTRAPS

**CALTROPS**

Webster's International Dictionary (2nd edition) defines a caltrop as "an instrument with four iron points so disposed that any three of them being on the ground, the other projects upward, used to impede the progress of an enemy's cavalry, etc." This definition is outdated now, as caltrops are no longer restricted to having four points, the points are not necessarily made of iron, and cavalry has become scarce. Nevertheless, the idea of an upward-projecting point upon which the victim impales himself is still very much alive.

The caltrop collection having the largest number of variations comes from Vietnam where they were used as antipersonnel weapons. In addition, a conventional four-point item has been received from Brazil where it had been placed on a street to puncture an automobile tire.

On the following page is a sample of a "new look" caltrop from Vietnam which follows the

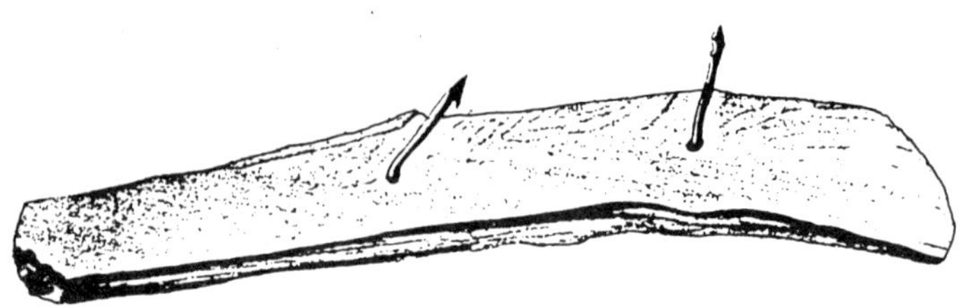

spirit, if not the intent, of Webster's definition. The device consists of a piece of board with two large nails driven through it, and if the number of points projecting upward have anything to do with it, it should be twice as effective as its conventional predecessor. To compound the problem of disengaging the victim from these points, the maker has supplied them with barbs. The Viet Cong, who operated in South Vietnam, made these devices with anywhere from one to seven points, each of which ranged from 2 to 12 inches long. To complicate matters further, they occasionally smeared the points with fecal matter or any other filth which happened to be at hand.

An improvement on the previous item, shown below, is the spike portion of a caltrop minus the

support board. The improvement consists of a second barb pointed in the opposite direction from that of the first barb, and so located that both barbs will lodge within the victim's foot. This arrangement increases the attending doctor's problem in removing the barb by preventing him from cutting the non-barbed end and pulling the spike out with a pair of pliers, not to mention the effect on the patient.

All these devices can be built just by looking at the illustrations; no complicated explanations are needed. With a little imagination the variations that could be built are endless.

## AUTOMOBILE TIRE DEFLATOR CALTROP

One such variation is the caltrop used to instantly deflate automobile tires.

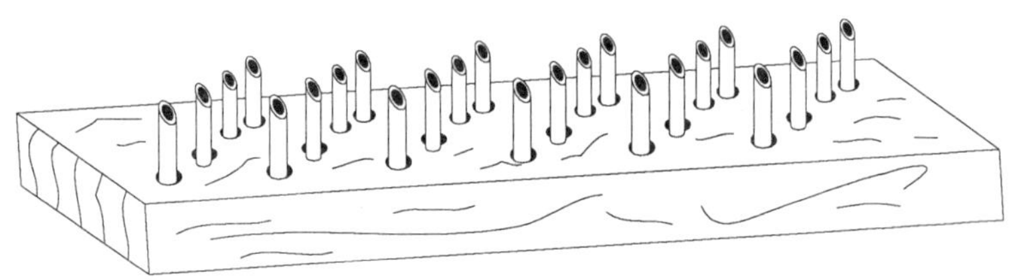

### Material Required:

Wood Board
⅛ inch diameter steel tubing
Hack saw
Hand drill

### Preparation:

1. Drill several rows of evenly spaced 5/16 inch diameter holes diagonally across the entire wooden board. The holes should extend into the board ¾ inch.

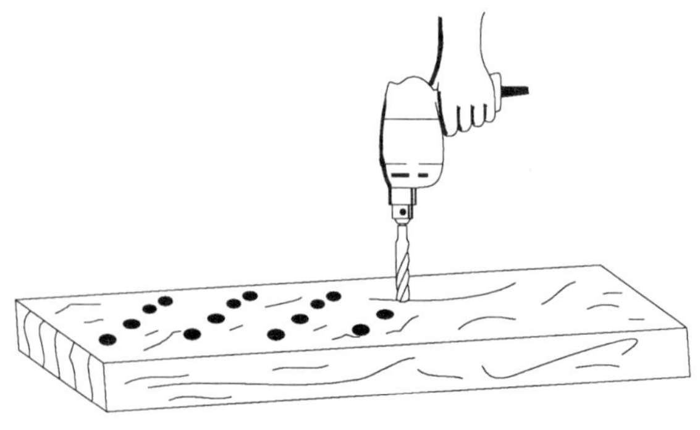

Wooden Board

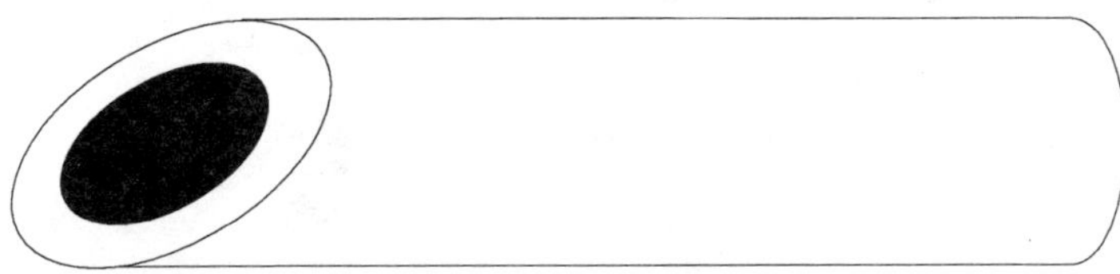

2. Cut the steel tubing at an angle to create a sharp point at one end and straight at the other end. Each should be approximately 1¾ inch long.

3. Place one steel tube in each hole in the wooden board with the pointed end up.

**How To Use:**
1. Place board across road or trail where vehicle will pass.

2. Secure board with weights at each end if necessary.

When vehicle passes, its tire(s) are punctured by the steel tubes. The air inside the tire passes through the hole in the steel tube instantly deflating the tire.

## CARTRIDGE TRAP

Four simple and easily obtainable components make up this mine.

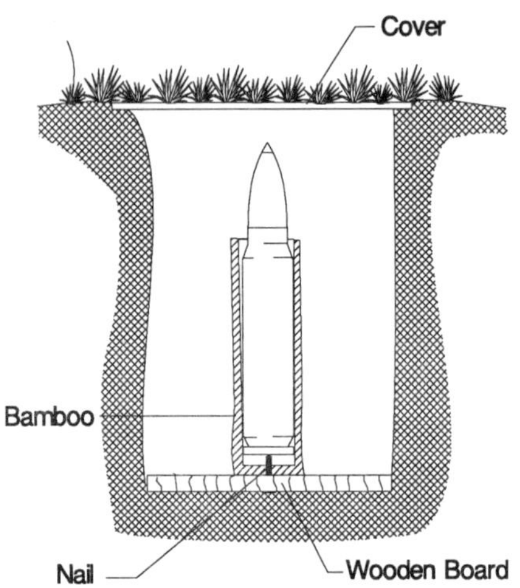


## Material Required

Bamboo tube
Nail
A piece of wood
Any small arms ammunition or M79 round
Hammer

## Preparation

1. Clean out the bamboo so that the ammunition fits snugly but has free movement inside.

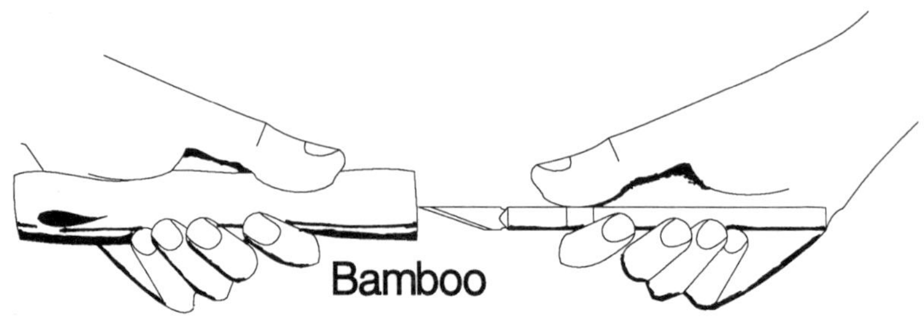

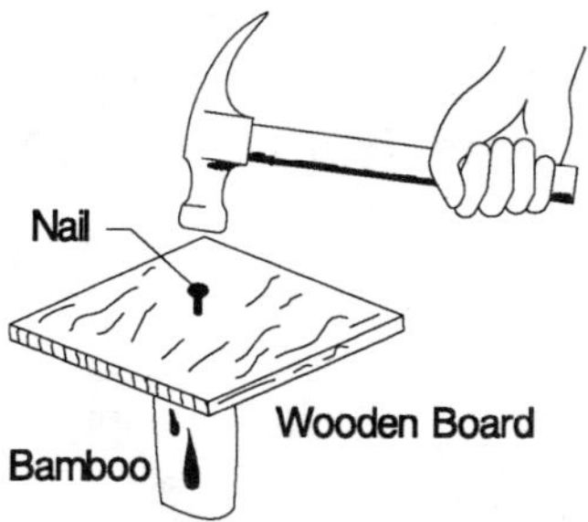



2. The piece of wood is used as a base. Place the bamboo tube upright on the wooden base and drive the nail up through the wood and into the bottom of the bamboo section. The nail must be centered in the bamboo tube.

3. Insert the cartridge into the bamboo so that the primer is touching the point of the nail.

**How To Use:**

Partially buried along a trail or path, the pressure of a man's foot stepping on the nose of the cartridge forces the primer onto the nail, firing the cartridge.

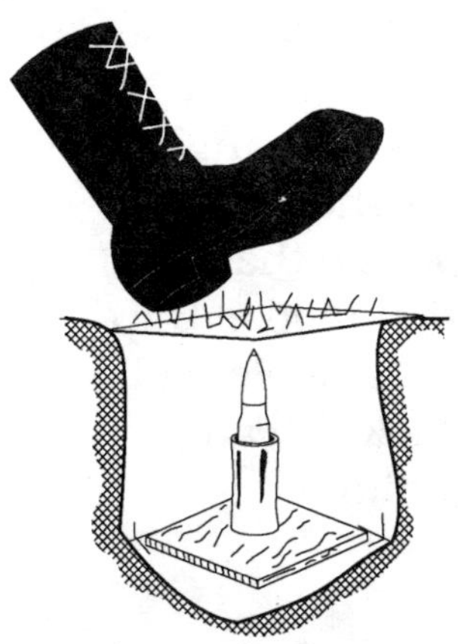

## MOUSE TRAP BOOBYTRAP

The Mouse Trap boobytrap is perhaps the most effective, easy, and deadly method of boobytrapping.

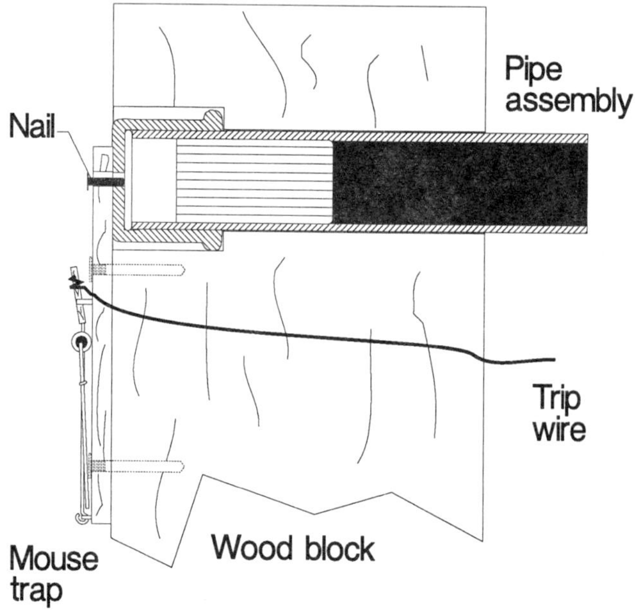


## Materials Required:

Steel or heavy galvanized pipe, 10 to 12 inch long threaded on one end. A 12 or 20 gauge shotgun shell must fit snugly inside this pipe but the brass rim should not.

A threaded pipe cap that will fit the pipe.

A regular wooden mouse trap.

A large flat head shingle nail.

Strong, thin black thread.

Hand drill and bits

Hammer

Knife

## Preparation:

1. Drill a hole through wherever the trap will be mounted. The pipe should fit snugly but be able to be inserted and removed easily. The hole must be larger at one end to accommodate the pipe cap.

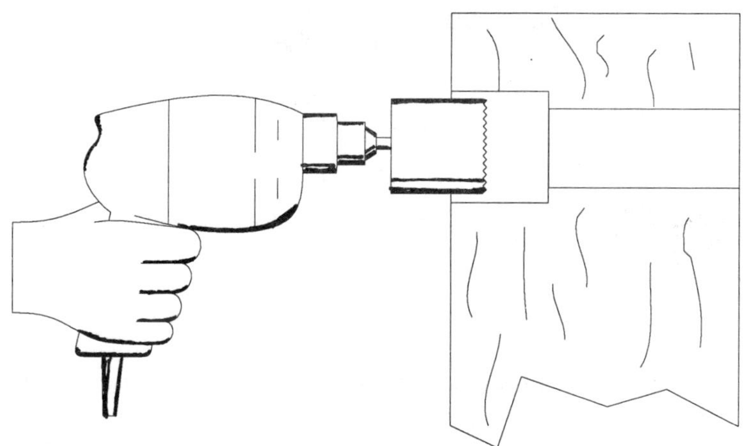

2. Drill a ⅛ inch hole in center of pipe cap.

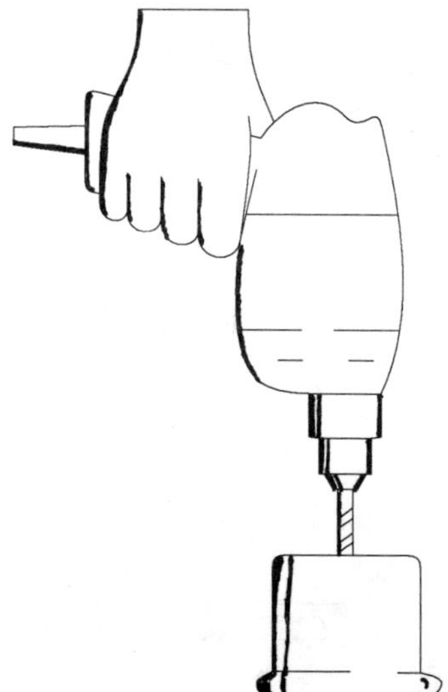

3. Insert the shotgun shell into the pipe and screw the cap tightly onto the pipe. The hole in the pipe cap must be directly over the primer of the shotgun shell.

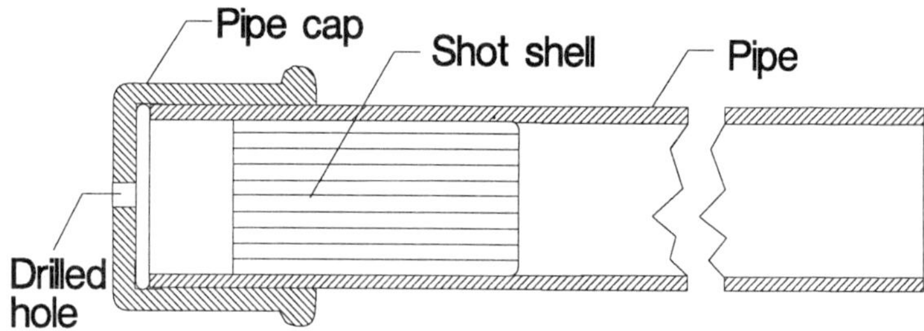


4. Drill a ⅛ inch hole in center of mouse trap in the area where the striker will hit the base.

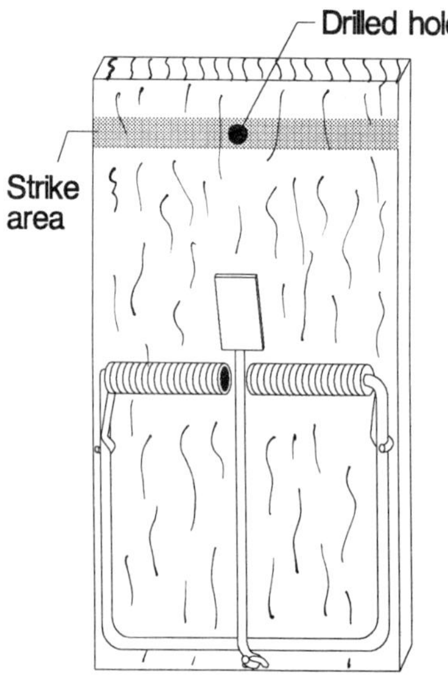

5. The pipe is then inserted through the hole in the mount and the mouse trap is nailed to the mounting surface in back of the pipe. The hole in the base of the mouse trap must line up with the hole in the pipe cap.

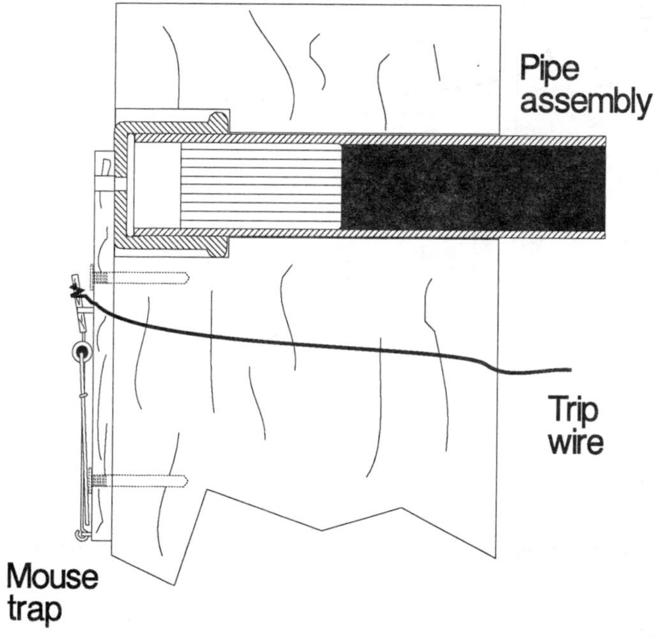


**To set the trap.**

1. Tie the thin thread to the trigger pan of the mouse trap and place across trail about 2 feet above the ground. Anchor on the other side of the trail. Do not secure the thread too taunt, leave some slack in it.

**CAUTION: This trap is extremely sensitive. Make sure nothing touches the trip wire or the mouse trap.**

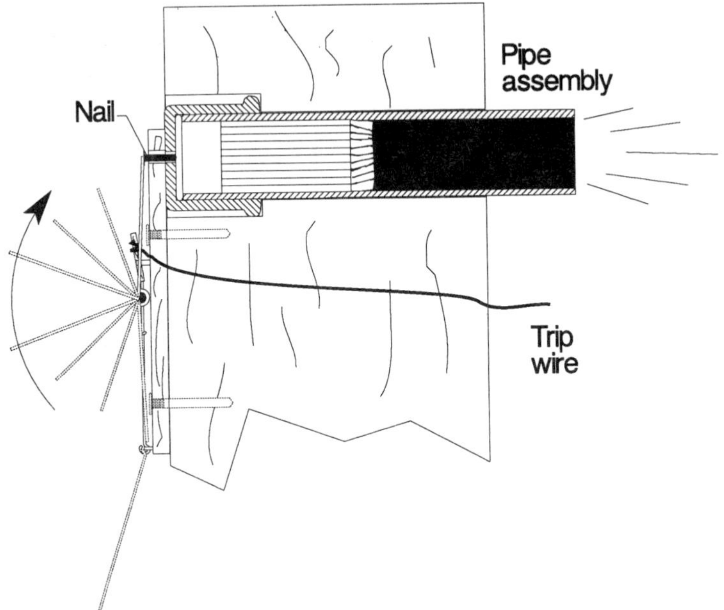


2. Stand behind and to one side of the trap and set the mouse trap.

3. Insert shingle nail into the hole through the mouse trap and pipe cap until you feel it against the primer of the shotgun shell. The trap is armed.

*NOTE: Always approach this trap from behind and remove the shingle nail to disarm it. Only place the shingle nail in the trap when you are ready to arm it. Use the pipes only once.*

This trap can also be set to activate when doors, cabinet covers, and appliances are opened. A great variety of triggering devices can be improvised to activate this trap.

# CHAPTER 4
# WEAPONS

## PIPE PISTOL FOR .22 CALIBER AMMUNITION LONG OR SHORT CARTRIDGE

A .22 Caliber pistol can be made from ⅛ inch nominal diameter extra heavy, steel gas or water pipe and fittings. Lethal range is approximately 33 yards.

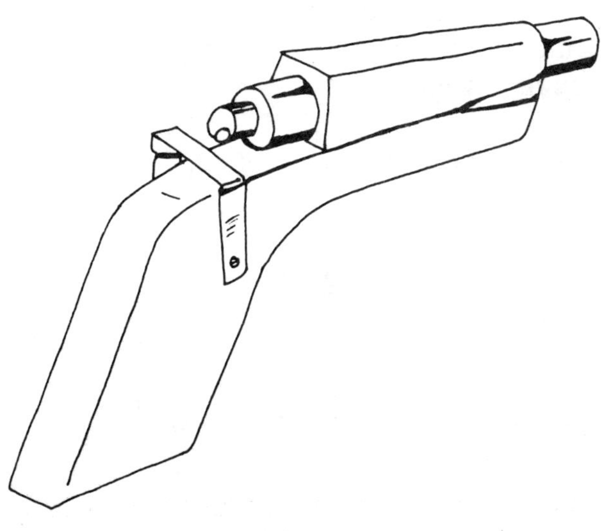

## Material Required:

Steel pipe, extra heavy, ⅛ inch nominal diameter and 6 inch long with threaded ends (nipple).

Solid pipe plug, ⅛ inch nominal diameter.

2 steel pipe couplings, ⅛ inch nominal diameter.

Metal strap, approximately ⅛ inch x ¼ inch x 5 inches.

Elastic bands.

Flat head nail—6D or 8D (approximately 1⁄16 inch diameter.)

2 wood screws, #8.

Hard wood, 8 inch x 5 inch x 1 inch.

Drill.

Wood or metal rod, ⅛ inch diameter and 8 inches long.

Saw or knife.

## Preparation:

1. Carefully inspect pipe and fittings. Make sure that there are NO cracks or other flaws in the pipe or fittings.

2. Check inside diameter of pipe using a .22 caliber cartridge, long or short, as a gauge. The bullet should fit closely into the pipe without forcing, but the cartridge case SHOULD NOT fit into the pipe.

3. Outside diameter of pipe MUST NOT BE less than 1½ times the bullet diameter.

4. Drill a 15⁄64 inch diameter hole 9⁄16 inch deep in pipe for long cartridge. (If a short cartridge is

used, drill hole ⅜ inch deep). When a cartridge is inserted into the pipe, the shoulder of the case should butt against the end of the pipe.

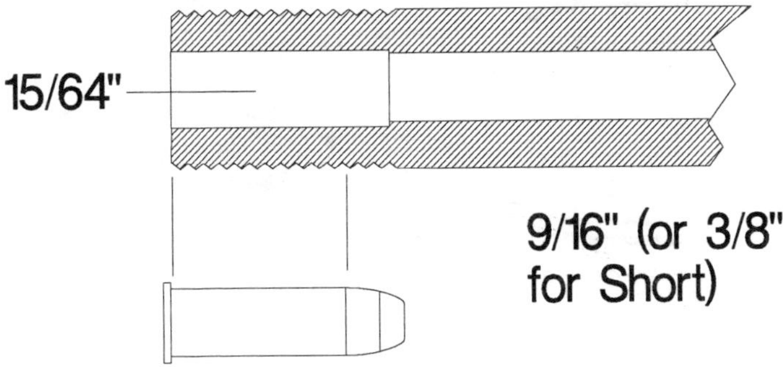


5. Screw the coupling onto the pipe. Cut coupling length to allow pipe plug to thread in pipe flush against the cartridge case.

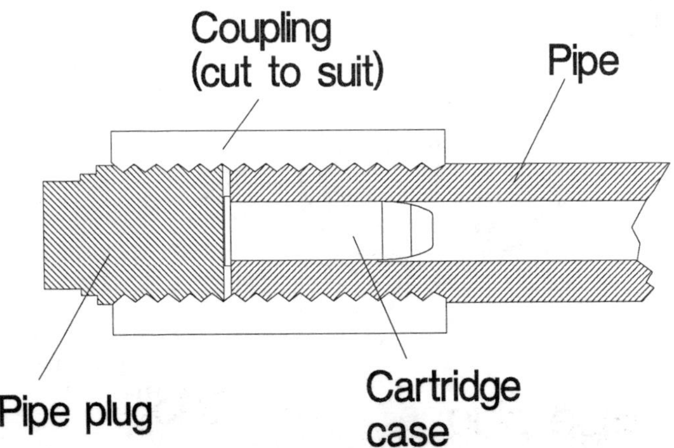

6. Drill a hole off center of the pipe plug just large enough for the nail to fit through.

*NOTE: Drilled hole MUST BE OFF CENTER in plug.*

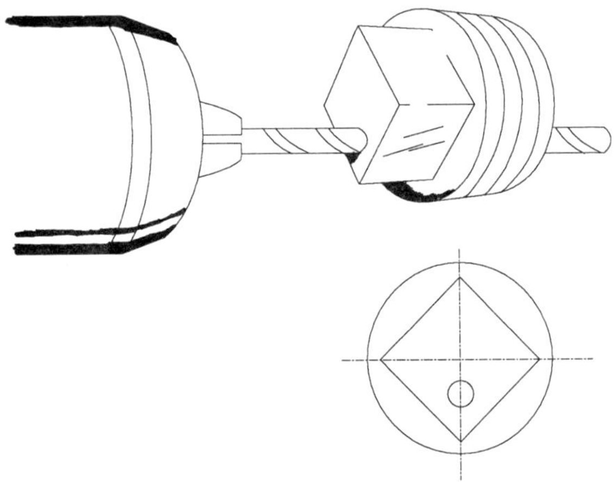

7. Push nail through pipe plug until head of nail is flush with square end. Cut nail off at other end ¹⁄₁₆ inch away from plug. Round off end with file.

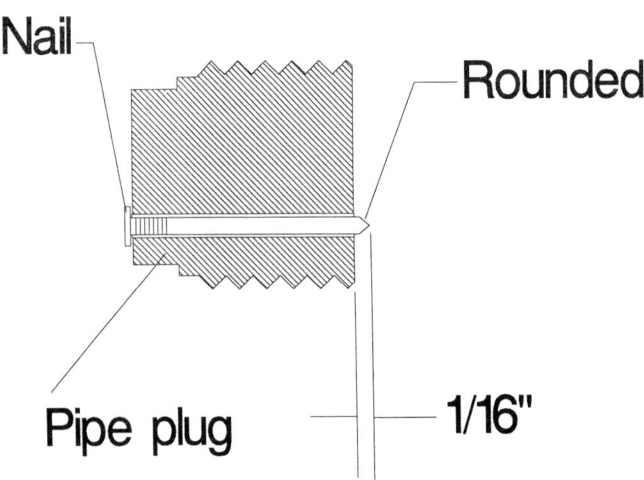

8. Bend metal strap to "U" shape and drill holes for wood screws. File two small notches at top.

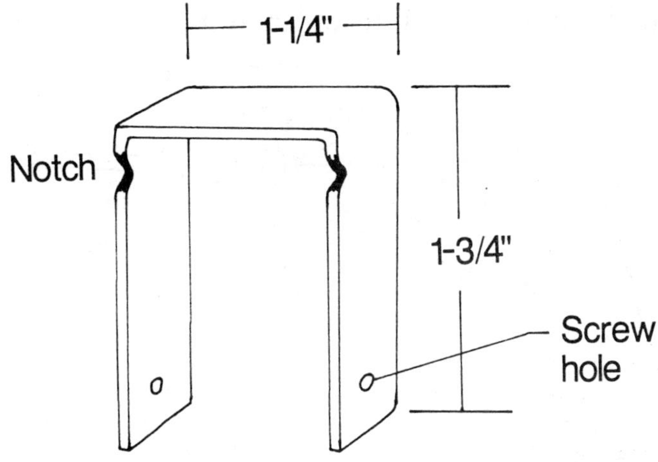


9. Saw or otherwise shape 1 inch thick hard wood into stock.

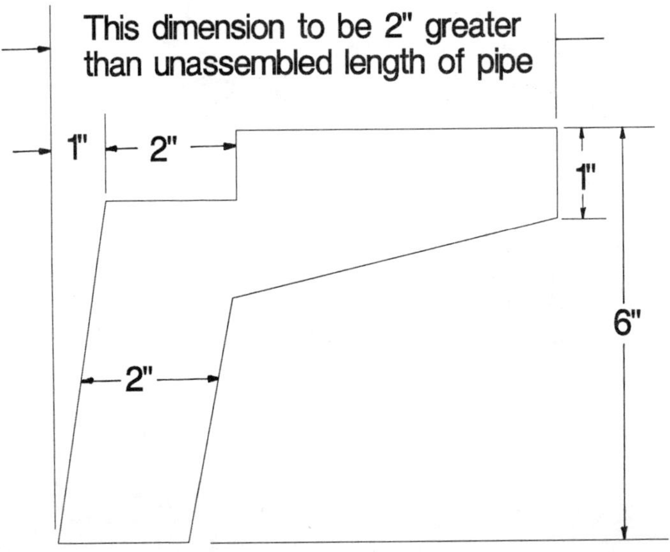

10. Drill a ⁹⁄16 inch diameter hole through the stock. The center of the hole should be approximately ½ inch from the top.

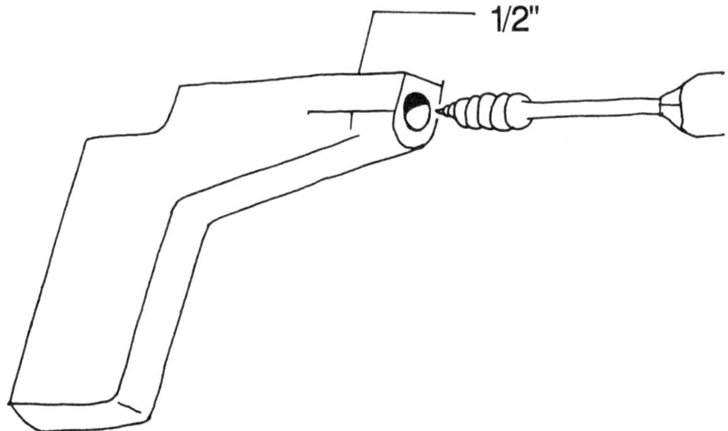


11. Slide the pipe through this hole and attach from coupling. Screw drilled plug into rear coupling.

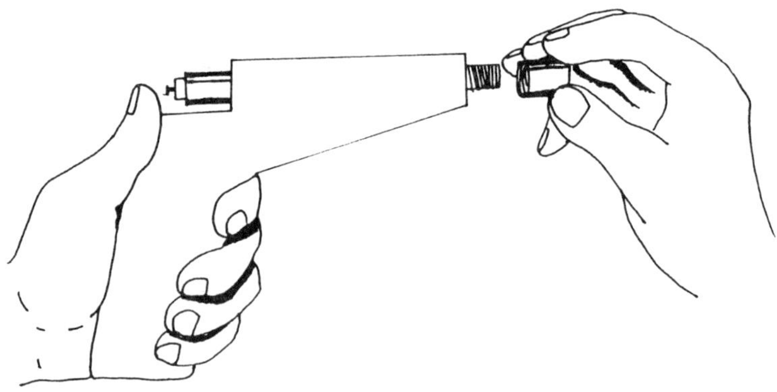

*NOTE: If 9/16 inch drill is not available cut a "V" groove in the top of the stock and tape securely in place.*

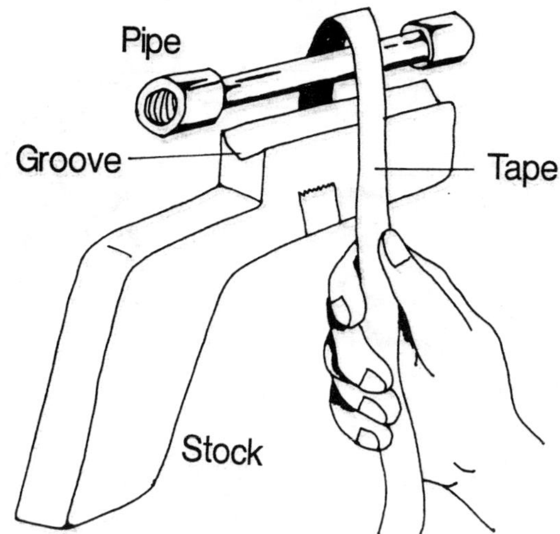


12. Position metal strap on stock so that top will hit the head of the nail. Attach to stock with wood screw on each side.

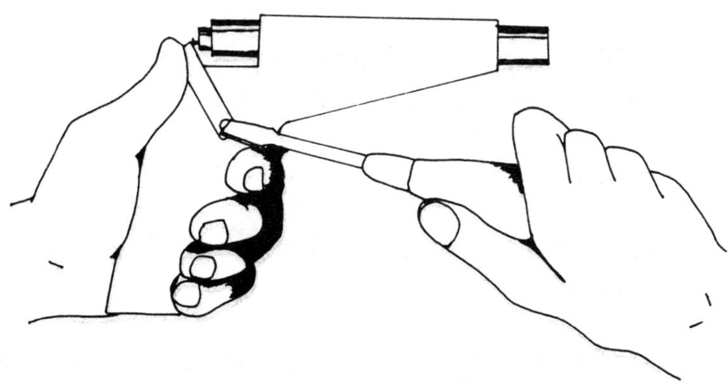

13. String elastic bands from front coupling to notch on each side of the trap.

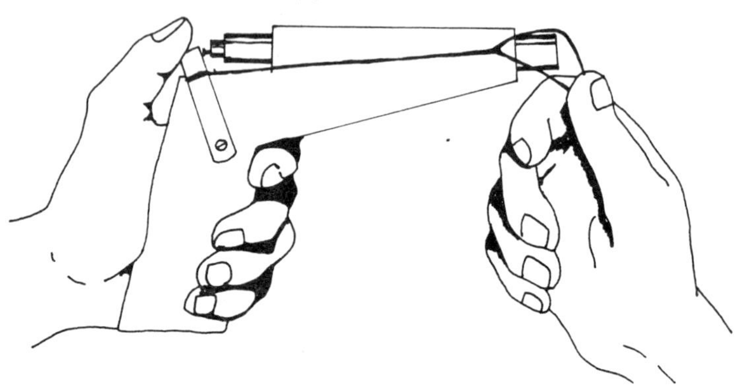

## Safety Check - Test Fire Pistol Before Hand Firing

1. Locate a barrier such as a stone wall or large tree which you can stand behind in case the pistol ruptures when fired.

2. Mount pistol solidly to a table or other rigid support at least ten feet in front of the barrier.

3. Attach a cord to the firing strap on the pistol.

4. Holding the other end of the cord, go behind the barrier.

5. Pull the cord so that the firing strap is held back.

6. Release the cord to fire the pistol. (If pistol

does not fire, shorten the elastic bands or increase their number.)

**IMPORTANT: Fire at least five rounds from behind the barrier and then re-inspect the pistol before you attempt to hand fire it.**

## How To Operate Pistol

To Load

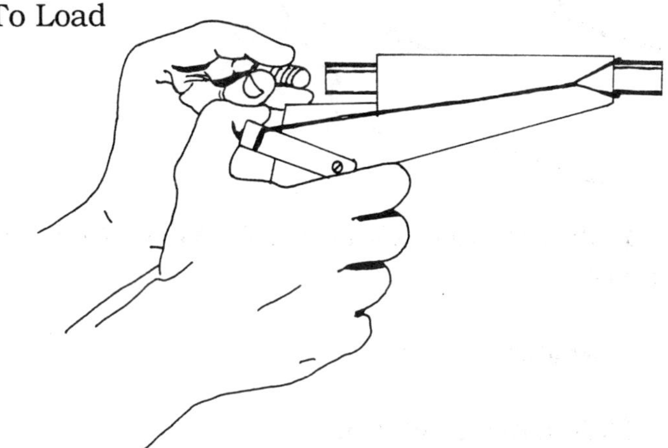

1. Remove plug from rear coupling.

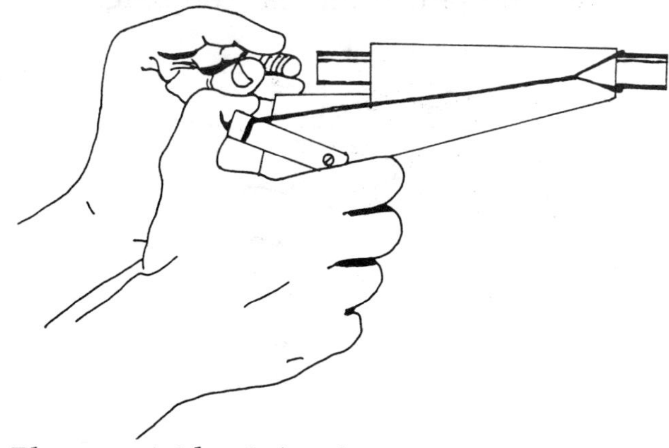

2. Place cartridge into pipe.

3. Replace plug

To Fire

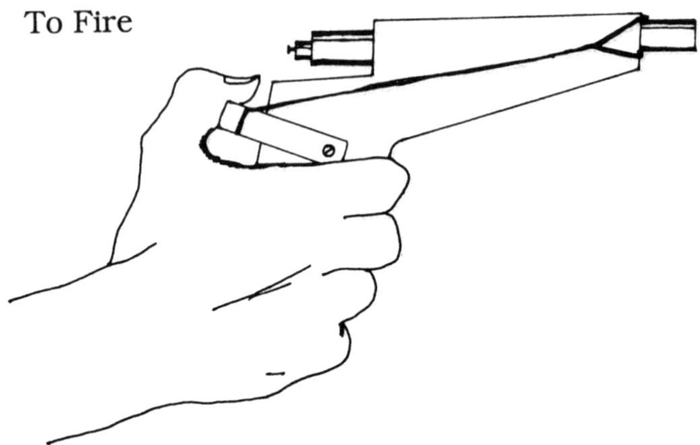

1. Pull strap back and hold with thumb until ready.

2. Release strap.

To Remove Shell Case

1. Remove plug from rear coupling.

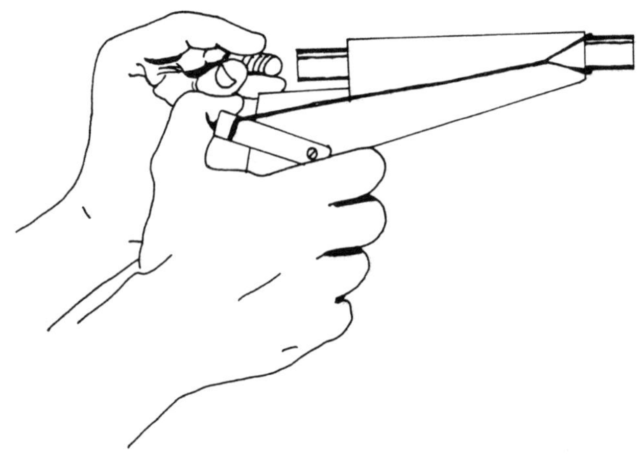

2. Insert steel or wooden rod into front of pistol and push shell case out.

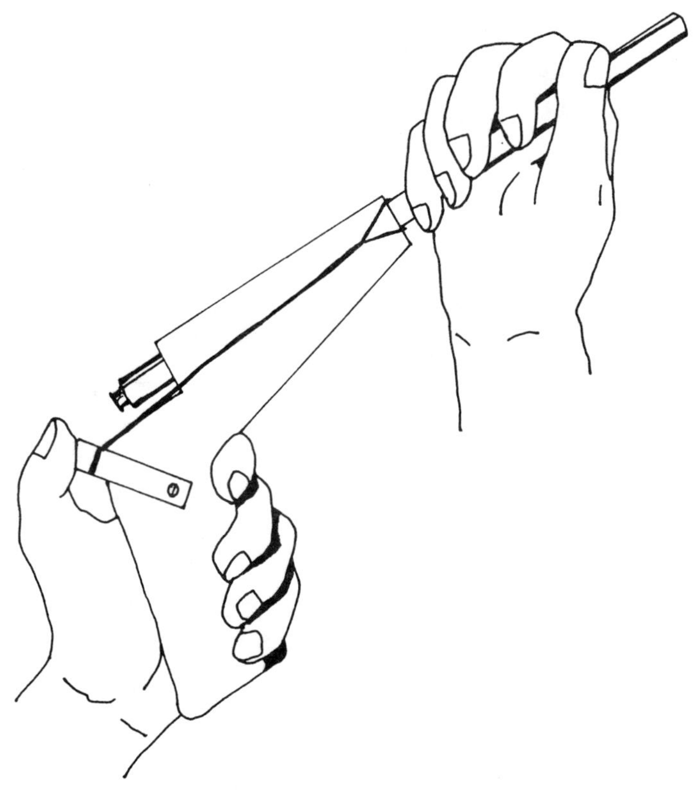

# OTHER TITLES

# VIDEO

**☐ THE BATTLE OF KHE SANH and MEN WITH GREEN FACES**

Two films on one videocassette!

**The Battle of Khe Sanh** provides an exciting account of the epic 77-day struggle for Khe Sanh—this siege, which was part of the most powerfully coordinated enemy offensive of the Vietnam war, was broken by a nearly unprecedented display of air power. This film depicts the courage displayed by the main defenders of the area—the 26th Marine Regiment.

**PLUS: Men with Green Faces:** Action-packed film on the U.S. Navy's elite fighting force—the SEALs—see how they prepare for and conduct unconventional warfare operations; parachute-jump training, jungle patrols, demolitions, and hit-and-run attacks. 59 minutes, Color.

**No. V18**                                   **$19.95**

**☐ RANGER TRAINING**

This interesting program presents the complete training program every Ranger hopeful must pass before he can wear the Ranger patch.

From demolitions and boobytraps, to survival and weapons training; every single subject must be mastered under actual field conditions. Other subjects which must be learned include rappelling, counter-guerrilla tactics, parachute jumping, and first-aid; all are included in this tape.

Also included in their entirety are two field exercises in which the Rangers are shown how to assault a communications station and blow it up, and how to conduct an airborne assault on an enemy camp; all in living color!35 minutes, Color.

**No. V58**                                   **$19.95**

**☐ SOMEONE SPECIAL**

This is the story of the Navy's Special Warfare Units. This film takes you into the world of mini-submarines, airborne assaults, and underwater demolitions. It depicts what it takes to become part of an Elite fighting unit such as a UDT or SEAL team. The full training program is presented including physical conditioning, underwater and ground demolitions, cast and recovery from rubber boats, reconnaissance, SCUBA diving, patrolling, and small unit tactics. 25 minutes, Color.

**No. V67**                                   **$19.95**

**☐ WEAPONS OF THE INFANTRY**

This action-packed color film provides the opportunity to see the following weapons in action: M-14, M-14A1 and M-16 rifles; M-60 and .50 caliber machine guns; M-79 grenade launcher; M-72 rocket; 3½ inch rocket launcher; 90mm and 106mm recoilless rifles; .50 caliber spotting gun, and 60mm and 81mm mortars.

Firing demonstrations showing the viewer how to operate every single weapon are included.21 minutes, Color.

**No. V4**                                   **$19.95**

**☐ THE M16 RIFLE: Operation & Functioning**

This film shows all features and capabilities of the M-16 weapon series; how loading, firing, and unloading is accomplished, as well as explaining the cycle of functioning for semi-automatic and fully automatic fire. 20 minutes, B&W.

**No. V3**                                   **$14.95**

**☐ BASIC MARKSMANSHIP—The .45 Caliber Pistol**

Correct aiming, gripping, stances, trigger control, and marksmanship principles are just a few of the techniques you can easily and quickly master through the use of this video program.

These practical, easy-to-follow techniques will turn you into an effective shooter, quickly and inexpensively.

15 minutes, Color.

**No. V16**                                   **$14.95**

## *30 DAY MONEY-BACK GUARANTEE*

Please send me the video(s) I have checked above. I am enclosing a check, money order or VISA or MasterCard information for $_________ plus $3.00 for shipping. For immediate service, call **TOLL FREE 1-800-472-2388**. (credit card orders only.)

☐ Please send me a FREE catalog (no purchase necessary)

CARD NO.________________________________________EXP. DATE_________

NAME________________________________________________________

ADDRESS_____________________________________________________

CITY________________________________ST_______ZIP___________

Mail to: **J. FLORES PUBLICATIONS PO Box 830131, Dept AH3, Miami, FL 33283**